In Search of Beauty

and

Other Poems

Second Edition

---

by

Winston Ka-Sun Chu
徐 嘉 慎

In Search of Beauty

This book was originally published in 2002.
This Second Edition is published in 2023.

To my Grandson Marius

To enhance

His enjoyment of poetry

INTRODUCTION

I have known for a long time that in the midst of a busy and highly successful professional career Winston wrote poetry. I know how difficult that is, as I have tried to do it myself, but the effort that it takes and the pleasure that it gives are always worthwhile. It is a privilege that I have now been able to read some of his poems.

Professor John White
University College London
November 2001

Never to allow gradually the traffic
to smother
With noise and fog the flowering
of the spirit

From 'I think continually of those
who were truly great'

by Stephen Spender

PREFACE

This selection of poems is printed at the persuasion of those who are dear and near to me. If the reader should enjoy any one poem, one stanza or even just one line, this work would have served its purpose.

The work has been designed for easy reading and enjoyment. It is hoped thereby to reach a wider readership. Literary merit, if any, is accidental.

The traditional style is adopted with some experimentation as rhyme and rhythm are useful means to generate images and emotions. The universal and ageless attraction of poetry in contrast to prose must be due, not only to the fascination of the ideas and imagery conveyed by its words, but also the allure of the sounds of its construction.

Winston Ka-Sun Chu (徐嘉慎)
1st January 2003
Hong Kong

PREFACE TO THE SECOND EDITION

The First Edition of this book of poetry was privately published in 2002. Surprisingly it was well received. Due to such encouragement, this Second Edition is produced with some editing and includes fourteen new poems and one translation of a popular short, simple but beautiful and timeless Chinese poem "夜思" by the famous poet Li Bai (李白).

The eternal hunger for art is because it touches the emotion of the heart, stimulates the imagination of the intellect and nurtures the sense of beauty of the soul. Though art may take many forms, yet to survive the test of time, perhaps it is best achieved through simplicity and clarity. As for poetry, perhaps the best achievement is from its enjoyment by its readers. This Chinese poem provides an excellent example and it is this style I try to emulate.

I am grateful to my son Dr. Chester Chee Wing Chu (徐志榮) for providing the photographs taken by him for this Second Edition which celebrates the magic of life and beauty.

Winston Ka-Sun Chu (徐嘉慎)
1st January 2022
Hong Kong

CONTENTS

Part IV – To Life

Part V – Last Words

Appendix

Part I

To Beauty

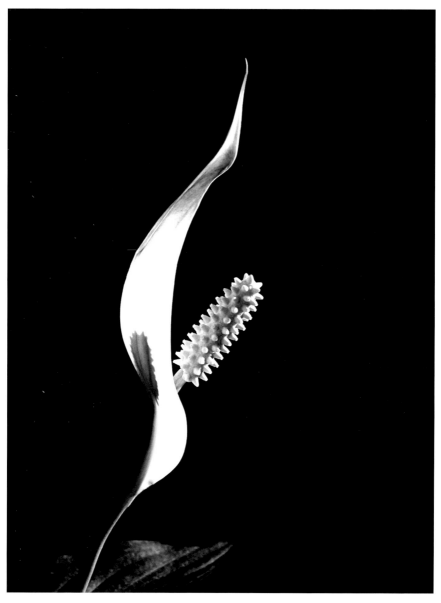

Peace Lily

IN SEARCH OF BEAUTY

Tell me where the lilies grow
Where the willows sweep the snow
Where the blue wings cut the sky
Where the mountains brood and sigh
Where the river runs its way
To itself would softly say:

"I see babies bright and gay
In the arms of mother lay
Change to youths so tall and fair
Stout hearts ready for their share
Ripples upon ripples roll
Life flows swiftly to its goal"

Sunrise paints the darkness red
Lines the clouds a silver thread
Lost in cradles of the mind
Beauty is for us to find
Struggling against worldly tides
Let truth and faith be our guides

Peacock Butterfly on Buddleia

PASSION FOR BEAUTY

Passion for beauty ever lives
In precious memories that seem
Haunting fantasies from a dream
Divine providence to life gives

Passion for beauty ever free
To ennoble humanity
To cherish for eternity
Loveliness just the heart can see

Passion for beauty never dies
But ever feeds the inner soul
To make the broken spirits whole
And life ever a paradise

Moon Rise over Beach

FULL MOON AT SEA

Silver galleon of the sky
Cloudy darkness sailing by
What mystery in the starless night
Paints the wonder of your light

Beneath a glowing canopy
Endless heaven endless sea
Dark waves twinkle silvery white
By the splendour of your light

Ever you set the spirit free
To indulge in fantasy
Entranced on this magical night
By the romance of your light

Flying into Sunset

ON AN AEROPLANE

To ride a stately aeroplane
On an ocean of cotton cloud
Blanketing by its milky shroud
Green mountains on a chequered plain

Now citadels rise tall and grand
Range upon range of frozen crests
Glaciers nurse their snowy breasts
Frost the land with an icy hand

Shiny ribbons of melted snow
From rocky hills and wooded dales
Meander down the verdant vales
Where streams through grassy meadows flow

Crimson iris of heaven's eye
Sinking below a fiery line
Paints ever changing hues divine
A kaleidoscope in the sky

In Search of Beauty

Alps from the Air

Then watch a giant octopus
Long tentacles of neon light
Stretch out into an empty night
Turning the dark world luminous

To a colourful galaxy
Of sparkling iridescent glass
On this metropolis of stars
Lands the heavenly argosy

Yosemite

THE MOUNTAIN

The night drops her umbrageous crown
And walks the western path away
I sense beyond your darkened shape
The awakening of the day

High in the air your lofty peak
Is sparkling in a field of snow
And in the coolest highland lakes
The twin sun rises from below

The mist uncovers hill and dale
That by its moving veil concealed
Now in the clear unfailing light
Your splendid image hangs revealed

With wind upon your ferny hair
And song birds in your leafy tress
And thundering milky cataracts
All gushing down your greenwood dress

Nevada

And clouds billowing from afar
Circle a halo round your head
You blazon near a hundred shades
Just when the evening sun has fled

Then when the sky's a shower of stars
Night pearls adorn the lotus bed
You look towards the eastern arc
And plan the stirring day ahead

You soar above the sea and plains
Mounted on a commanding throne
You fill the sky, you hold the wind
You rule the green world all alone

In Search of Beauty

Rainbow

THE RAINBOW

With wonder and with leaping heart
I trace the rainbow to its start
And there I find a beauteous world
Of light and colours richly curled

You bridge the mountains, span the sky
A band of ribbons dazzling high
That arches with surpassing grace
From heaven's gorgeous treasure case

The red of ruby glint of gold
The blue of sapphire deep and bold
From amethyst a purple shade
And green of flawless Chinese jade

When caught in sorrow's strangling rope
I find in you a source of hope
Ascending your ethereal stair
I brim with joy and ban despair

To glory and to wisdom march
Worthy of your triumphant arch
Taking the colours from your gown
I wear them in a conqueror's crown

Part II

To Nature

Spring Blossoms

ODE TO NATURE

I only want the gentle kiss of
    spring
To feel the passionate hand of the
    summer sun
To dance with the leaves upon
    autumn's wing
And play with winter's white innocence
    just for fun

To blend my wanton spirit with
    the seasons
And follow Nature to her
    secret lair
There to confess my love and all
    the reasons
I keep her forever my mistress fair

Pink Tulips

LEAVES IN SPRING

Fresh young buds smile to welcome spring
Fair messenger from Nature's King
That sleep is over light is born
Arise to greet the radiant morn

So they dance while the sky lark sings
Whose music through the gay air rings
Dainty wings with emerald glow
They flutter when the breezes blow
Like young fledglings they yearn to fly
Headlong into the deep blue sky

When spring in showery mantle wept
Drops of tears down their faces crept
To weep and weep in rainy grip
Till sunshine peeps from the dark cloud's lip

At dawn they would the sweet dew drink
Wondering why the sky's so pink
They stare into the darkened height
And woo the shining stars at night
They care not what their fates will bring
They only know that it is spring

Dew Drops

SPRING RAIN

Freshest of surprises, unexpected guest!
All of nature you dress in their loveliest
The greenest foliage, soft and imbued with light
Rivals the sky-blue and scorns the lily-white
Adorning with crystal drops and shiny coat
You carry sweet moisture to the forest throat
To drink and by drinking be thirstier still;
Only the new buds are gulping to their fill
The honey-making of wild flowers will cease –
Rain-drenched the pollinated wings of the bees

Then to the towering city you nobly stride;
Splash on the roof-top and tap the window side
A sudden deluge causing a civic fuss
You hitch a pattering ride on the autobus
The shallow gutter brims with a busy stream;
Umbrellas and raincoats strike a working team
The air is tense – sullen sky and streaking lines
Have sate the city drunk deep in new-brewed wines
With crowds escaping from your watery clutch
You freshen the landscape with a magic touch

In Search of Beauty

Back Street at Night

THOUGHTS ON A SUMMER NIGHT

For it is a wild wild night
Warm breeze lures me from my bed
Rich folks doze in golden dreams
Poor men fret their airless shed

For it is a wild wild night
Lights are dancing on the sea
Rich folks burn their lanterns bright
Poor men save their candle fee

For it is a wild wild night
Insects chirp a lusty strain
Rich folks praise it as divine
Poor men suffer silent pain

For it is a wild wild night
Trees are greening night air pure
Rich folks draw their sweetness all
Poor men smell the garbage sewer

In Search of Beauty

Sunset at Yacht Club

For it is a wild wild night
Jewels twinkle in the sky
Rich folks love a starry night
Poor men sleep once in the dry

For it is a wild wild night
Ferries crawl a gleaming path
Rich folks like to midnight bathe
When did poor men have a bath?

For it is a wild wild night
Deep calm dwells upon the hills
Rich folks smoke a quiet pipe
Poor men's child makes hungry shrills

For it is a wild wild night
Fishing boats sail out to sea
Rich folks cry: "a wondrous sight!"
Poor men brave the stormy sea

For it is a wild wild night
Drifting into slumber-land
Rich and poor shall lay alike
Rest safe in Our Maker's hand

Japanese Maple

AUTUMN NOCTURNE

I hearken season of regrets
Sad music of your rattling leaf
Misgivings that your mood begets
Instil in me a deeper grief

Deep silence in yon silent street
And on the trees so pale and red
Such silence is so bitter-sweet
It seems to say the world is dead

Cold and wet the funeral bed
Of leaves in deep eternal sleep
And lit by street lamps overhead
With long sighs all the old oaks weep

The sycamore the first to cry
So melancholic yet serene
Its fallen tears have waxened dry
Line the pavement an earthen green

Colourful maple autumn's pride
Would swing her red robe to the breeze
Is wooed she falls as winter's bride
Blushes a deep red by its tease

Silver Birch

The lone birch brooded all night long
Its leaves would rustle just by chance
Breaking the silence into song
They fly to join the north wind's dance

Whipped high and low the whirling gusts
From mountains and from northern plains
They chill and thrill with icy thrusts
Tainting the grass with frosty stains

The moon is frozen in the sky
And frozen every distant star
Last week the gurgling brook ran dry
And left behind a barren scar

Dark streets the autumn night embalms
In wind in silence and in rain
And lulled on yonder leafy arms
The young birds twitter all in vain

The city deep in slumber heaves
And still the grey mist broods on height
Westward the darkness draws and leaves
The heavens a fish-belly white

Sunset Sky

So creeps the night yet nearer dawn
A dozen rapiers pierce the sky
Through purple clouds the rays adorn
And trim the darkness by and by

A symphony of colours lights
Up the wakening russet world
Now in the eastern mountain heights
A golden banner is unfurled

Snow Covered Branches over Water

WINTER TWILIGHT

Cold stars are watchful in the sky
While all the whirling leaves rush by
The river babbling in the weeds
Has sown their melancholy seeds
And night owls hooting through the night
Are dreading winter's frosty plight
At dawn the sky's an eerie glow
The first north wind begins to blow
Autumn mellowed and now must go
Winter twilight will foster snow
Softly treads the departing night
Shadowed by rousing birds in flight
Bleak sounds of footsteps leaving home
Echo across the icy tomb
The day breaks faint the last star burns
A deep peace spreads the river churns
The moody mist on trees still lush
Is lifting in the twilight hush

Leaves and Shadows

I WALK IN SUNSHINE

I walk in sunshine
Far from the shadow
A moment divine
In Nature's cradle

Far from the voices
That cry out in pain
My heart rejoices
Let them cry in vain

The blossoming flower
Knows not pain nor grieves
Hears from hour to hour
Music of the leaves

Safe in Nature's arms
My sun is shining
When life lost its charms
I join the dying

Red Admiral Butterfly on Roses

ONE WITH NATURE

Carefree playful butterflies
Frolic in a summer sky
Rapt in their own paradise
Let the mad world pass them by

May melody of the lark
Resound of hope through the air
Carry comfort to the dark
Lonely abyss of despair

Boundless ambitions of man
Reach across the universe
Yet praise the Lord's deeper plan
In the rhythm of a verse

Man though nature's favoured child
Still one with the sea and sky
Near kin of all creatures wild
Humble in the Maker's eye

Part III

To Places

War Memorial

THE HEROES' GRAVE AT BUDAPEST

Do not grieve at the heroes' grave
Their dauntless spirit did not die
Still honoured the young lives they gave
When guns answered their freedom cry

Tanks charge the bullet-riddled wall
To extract the ultimate price
Blood-splashed the brave warriors fall
Yielding their final sacrifice

Now in the country of the free
Their blood colours the twilight sky
In the remembrance of the free
Thunder echoes their battle cry

Low Tide on Beach

PORTHPEAN BAY

On rolling seas a rocky seat
The roaring ocean at my feet
Far from the city noise and care
For this serenity to share

Green is the grass atop the cliffs
And green the shallow sea that gives
The crescent shore a verdant fringe
And the dark reefs a deeper tinge

The rugged coast a sweeping curve
Of sunken rocks and torn-out earth
It owns the solitude I seek
Concealed in secret cove and creek

Wave-washed spreads the glistening sand
Bold battlefield of sea and land
Like white-maned stallions rollers arch
Or white-plumed warriors on the march

Rock Pool

Again and yet again it rolls
The raging surf against the shoals
Each running ridge of dazzling foam
From far away where wild winds roam

The screaming gulls from cliff tops hurled
Half mad in crazy circles whirled
The screams that they in terror cried
Bounced off the cliffs from side to side

Now high on rocky ledges perched
Their eagle eyes the blue waves searched
And in a sudden twist one soars
Its victim crushed by iron claws

Among the rocks in limpid pools
The seaweed lies upon the ooze
And limpets half-baked by the sun
Await the stars when night has won

The hissing froth the shooting spray
The glittering ripples on the bay
Will still be here when I am gone
Still praised by others yet unborn

Waterfall

NIAGARA

I hear thunder in the wind
That roars with a roaring deep
Soars the spirit from restless sleep
With rhythm fierce but not unkind

Rainbows reach down from the cloud
Which the leaping fumes do feed
The vow to score a mighty deed
Rings in my captive mind aloud

On on gushing whirling waves
Down a giant wall to pour
The power of your living awe
Ever my guarded soul enslaves

Still it flies the silvery spray
From a battle with the rocks
A distant mist of chilly shocks
Comes with the dropping of the day

Rush Hour Traffic

TO HONG KONG AT NIGHT

Land of fascination, light and darkness
Add to the gleaming harbour one more stain!
The hanging stars that seem to be sparkless
Shine to enrich this fanciful domain
The sleeping hills are ridges of shadows;
The tall lean buildings glow like towers of light
I wander beyond your concrete meadows
And stray into a jungle draped by night

And there I hear such strange and fearful sounds:
Murderers' knife clash, screams of the haunted
Night vultures making their dangerous rounds
Thrilling sirens, sobs of the unwanted
Who answer destiny's unequal call
Tenaciously: and just like staunch willows
They bend with an unconquerable squall
And surviving, incarnate as heroes

Street Lamp over Stairs

So different from the harmonic strains
Played to dancing lights and moonlit waters
Exotic gardens' glamorous refrains
Merry-making from exquisite quarters
Of the elite who, too endowed with ease
Stand not fate's trial; but like rootless flowers
Collapsing with the slightest hostile breeze
Sink into existence's brutal bowels.

Land of fascination, tears and laughter
Add to the teeming city one more sound!
Recollections that come streaming after
Fill me with thoughts as I tramp homeward bound
The sleeping hills are masses of shadows;
The tall lean buildings flare like shafts of light
I walk again in your concrete meadows
But can't forget that jungle draped in night

In Search of Beauty

Derelict Wall

NIGHT ON THE BOSPHORUS

Silvery sea by an enchanted land
Where ancient minarets still nobly stand
The compassionate moon steadfastly shines
Though empires fall and even faith declines

Yon silence arouses a myriad sounds
Of awesome clamour from old battlegrounds
There conquerors next to the vanquished lay
As blood-drained carrion beside the bay

Wild grasses grow now on top crumpled walls
Faint stars gaze sadly upon roofless halls
Hot desert winds disperse the sultans' bones
Loud tourists degrade the emperors' thrones

Riches prestige all worldly pettiness
Are vain illusions in an emptiness
Timeless waves along the Bosphorus run
But human endeavours by time undone

Narcissus

NIGHT OVER VICTORIA HARBOUR

All is hushed – a sensuous fragrance
Floats across the rainbow sea
Bathed the night in moonlight radiance
Yields up all her charms to me

Spreads the sky a priceless treasure
Set with diamonds and with pearls
String the gems in stately gesture
On the night clouds' softest curls

High up streaks a shining comet
Naked flares its fiery spur
Falls towards a distant planet
Quick becomes a formless blur

Venus shows the loveliest lustre
Fringed with just a marine blue
All the sky a dazzling cluster
Sparkling like the morning dew

Orange Tulip

Hurls the moon ethereal lances
At a liquid plate of gold
Every shimmering ripple dances
For this dream-enchanted wold

Lines black hills a jewelled stairway
Banister of sapphirine haze
Leading up a misty gateway
Entrance to a starry maze

Like a gold flame on the mountain
Burning through a splendid shroud
Splits the gloom a glowing fountain
Luminous by a radiant cloud

Far to shore ten thousand lanterns
Glistening on the mountain slopes
Many are their coloured patterns
Liken the celestial globes

Crocosmia

Which of man and which of heaven
Several flash the winking Mars
Guide the eagle's man-made brethren
Comrades of the pilot stars

Sail from out a livid vapour
Posing on a winged lyre
Galleons of translucent amber
Plunge down earthward spitting fire

Towers a tribe of glittering Cyclops
Gazing with crystalline eyes
Blink aside their topaz tear-drops
Frozen in this paradise

Round yon distant dark abysm
Stands this colourful array
Steals the gay tints from a prism
Lines the bay a Milky Way

Japanese Iris

By the dark shadow of the pier
Hangs a row of neon suns
Shames the shy night to disappear
Rips her thin veil all at once

Calmly rest the ocean giants
Watch the night tide running full
To their old foe will defiance
In a haven colourful

Anchored here this best of seasons
As the night wind waltzes through
Ghastly glare their mast top beacons
Haunted through a ghostly hue

Glides a silent gleaming liner
A white satin trails her wake
Never sailed a sea queen finer
Sighs the sea-shore for her sake

Balloon Flowers

Casting light paths on the mirror
Quivering on its shifting glass
Bright as streams of molten silver
Blazing wild fire as they pass

To the bosom of the ocean
Tide-swept on her heaving breast
Sets the land end lights in motion
Ever climbing a climbing crest

Lustrous the night's darling navy
On a land-locked everglade
Netting their frail-finned quarry
To a star-dust serenade

Lights like fireflies on a quagmire
Hover in this timeless realm
Neath a sky of ancient sapphire
By their charms to overwhelm

Drink of the night's wind-brew vintage
Savour of the dancing brine
Spellbound by this rapturous image
Dream-struck in the full moonshine

Old Man Asleep

THOUGHTS ON OLD AGE WHILE AT SEA

Like darkness falling on the sea
Old age is creeping up on me
Grey shadows in the fading light
Evoke the terror of the night

No more the joy of golden days
But pain in parting of the ways
Manly powers' tragic decline
Warm tears flowing – quietly mine

Across tired waves the cold wind howls
As dark Death like a sea gull prowls
The longing for departed friends
Lends comfort when the journey ends

Fear of the mysterious deep
Breaks the lure of eternal sleep
Twilight upon the sea of time
Still strive to make the light sublime

Part IV

To Life

Bonfire

TO LIFE

Not for fortune not for fame
Just for sheer hell of the game
In the flicker of a flame

Transient this universe
Ere our atoms far disperse
Will a blessing not a curse

Helpless in life's stormy main
Bear the burden wear the chain
Face the prison drown the pain

Searching for a paradise
It's there where our true heart lies
By sun and by tearful skies

By our hunger for the right
Lift the blindness from the night
Light the wisdom of our sight

Warrior

Wean our vanity of lust
Ere the soft flesh melts to dust
Hail the karma of the just

Reach out for a starry goal
Seeking in our earthly role
Ennoblement of the soul

Nurture a spiritual light
Through life's tempest shining bright
May lost travellers share the light

Not a hero not a knave
Not a craven worldly slave
Live the fullness of the brave

Yellow Tulip

TO LOUISE

Amidst life's confusing chaos
Destiny happened to be kind
Thoughtfully brought us together
That I may so lovingly find
Your smile sweet as springtime roses
And heart pure as high mountain snow
Spirit with deep honest courage
And beauty of your poet soul
A brave journey we have travelled
Through the tempestuous tide of time
To share precious memories
Till the last winter rays of time
You will forever be cherished
We shall forever firmly stay
Yet hand in hand and soul with soul
On every lovely summer day
And every lonely autumn night
And if there is life hereafter
Yet you will still be dearly loved
In our many lives together

Squirrel

CHESTER

Is he an angel
Is he a toy?
An angel smiles
But not as sweet
A toy does not
Have such soft feet
He is a baby
A bundle of joy

Is he a sweetheart
Or a little boy?
Just see him smile
And fall in love
Hug him all day
But never enough
He is a darling
A bundle of joy

Boy Running on Beach

To set him laughing
And then enjoy
Those twinkling eyes
And teasing dimple
Contentment for me
Is just so simple
He is a blessing
A bundle of joy

Ten dainty fingers
He does employ
To tell the world
How to rejoice
At all the wonders
That it enjoys
He is my Chester
A bundle of joy

Ladybird

KAREN

Karen is my darling girl
Lovely as a precious pearl
Trace her twinkle to her nose
Shower her kisses from head to toes

See her tottering to and fro
Feel the warming heart aglow
Watch the mischief on her face
Her bright-eyed innocence, her ways

Smiles and laughter quick and sweet
Sometimes even such a treat:
Hug me tightly for a kiss
What contentment and what bliss!

Hear her singing her own words
Waving, talking to the birds
Calling for me all the time
"Daddy, Daddy" like a rhyme

Sea Urchin Shell on Beach

At times such a cutie pie
Quite the apple of my eye
And then naughty as can be
Tickles, giggles endlessly

Hold her tenderly in my arms
Stroke her gently with my palms
Join the sweetness of her dream
Floating softly we would seem

Take her firmly by the hand
Picking sea shells on the sand
Meet with nature by the sea
She, I and all in harmony

Resting Lambs

TO "MOUTON" MARIUS

Be kind that your heart will know virtue
Be gentle that your soul will find beauty
Be humble that you will earn respect
Be loving that your days will be filled with love
That a lifetime will not be wasted
Try to make every moment precious
Life is a fair and wise arbiter
It gives not what one desires but deserves
Its bounty cannot be owned but only felt
Close your heart and fists and people will fear you
For tight fists can show no welcome
But a warm embrace needs open arms and hands
A closed soul can merit no grace
And closed eyes and mind can see no light
Only goodness can bring the best of life
And blessings can only enter an open heart
For life is but a mirror
We see and receive what is inside us

Yellow Chrysanthemum

COME TO ME

My Love come to be with me
Then our days will ever be
Blest with joy and poetry

Profess my passion in a verse
Loving you in so many ways
I hold you in a fond embrace
To dream of our own universe

Though heartless the rhythm of time
Stanzas of love stay bitter-sweet
I lay my lyrics at your feet
Vow love eternal with this rhyme

Come to share this life with me
Then our world will always be
Blest with love and poetry

Mating Butterflies

MARRIAGE

"Marriage" says the Priest "is a sacrament"
He turns his Bible, quotes in merriment:
"Solemnly in the sight of God, we ought
To tie this saintly matrimonial knot."
The bridegroom sweeps the young bride off her feet;
Their beaming relatives each other greet
But quick the ceremonial garbs are shed
They gaily hurry to their nuptial bed

"Marriage" says the Registrar " is a rite.
From Section Four, the Civil Law I cite:
'A monogamous association
For life (or else until dissolution)'
All legal forms and moral norms are met
In legal dignity this seal I set."
He charged fifty dollars for his discourse
(Such is the fee for lawful intercourse)

Mating Butterflies

"Marriage" says the Merchant "is a venture,
An investment backed by an indenture.
My personal fortune it enhances
(The couple just have to take their chances).
Full twenty years of capital I've sunk;
Now dividends! Tonight I shall get drunk."
A property settlement they conclude;
The groom pays to hold the bride in the nude.

"Marriage" explains the young Bride "is a phase
Of eternal beauty, angelic grace.
We are a perfect union of the soul!
(To keep me happy is my husband's role).
He will make me a superlative mate,
To God and law this marriage vow I state:
'To love and honour, cherish forever
(This present of diamonds, gold and silver)'."

Mating Damselflies

"Marriage" claims the Groom "is a thing of pride,
Heroic measures, conquest of the bride.
We are a perfect union of the soul!
(To keep me happy is my new bride's role).
She will make me a superlative mate,
To God and law this marriage vow I state:
'To love and honour, cherish forever
(This sense of pride and a woman's favour)'."

"Marriage" quotes the Lawyer "a social fad
Of lawful union between love and bread.
The only danger of matrimony
Is the quantum of the alimony."
He gently questions the dejected spouse:
"Who takes the children? Who retains the house?
And do you want a divorce petition
Or just a judicial separation?"

Statue of a Judge

JUSTICE

Is it justice or just ice
That frosts the truth from the eyes
Of those trusting in the Law
Oblivious of its flaw

"Justice" a mere honoured word
More dishonoured than absurd
Balanced on a human guess
Armed with bias to assess

Guilt and innocence alike
Turn upon a fickle psyche
Cause the just man to despair
What is justice, what is fair?

Is it a mere legal game
Played with humans on a frame?
Mark the cunning legal sleuth
Woe the silent rape of truth

Sometimes wielded as a whip
For a political grip
Puts the Hitlers on their thrones
Screaming "Justice", breaking bones

Statue of Nike

"Justice only by my Book"
Cries the Christian with a look
Shouts the Moslem loudly too
And the Catholic and the Jew

Justice what its colour is
Not more wisely said than this;
"Black is evil, white is good"
Or "white is evil, black is good"

Justice a clear fervent call
For humanity to all
Equal despite race or creed
Measured only by the deed

Makes a human of the beast
Cuts his savage roots diseased
In the zenith of his mind
Plants salvation for mankind

In Search of Beauty

Chinese Vase

THE CHINA DREAM

I wish that I shall never wake
From this wistful rapturous dream
That the modern Middle Kingdom
Will follow the righteous way
To wield real power through wisdom
Earn respect by a noble heart
Trust worthy of a peaceful soul
And friendship from true helpful hands
To uplift the human spirit
Singing songs of innocent joy
Like a father to its people
Rich with quality and virtue
A strong force for eternal good
And a haven from pride and greed
A staunch bastion of freedom
A Utopia upon Earth
Standing tall before all nations
As inspiration to the world
Such a dream I pray that Heaven
Will make true till the end of time

Pink Magnolia

SILENT LOVE

Do I love you any the less
Though my love I do not protest
To love in silence to love best
The running wave not to possess
When long have I adored the sea
Called her your name and in her arms
To find your tenderness your charms
Both ever lovely ever free

Do I love you any the less
Though passion the long years subdue
To love unwavering to love true
The yearning heart not to suppress
To leave our imprint by the shore
On a bed of ever green grass
Timeless to watch the rollers pass
These are the memories to live for

Weathered Copper Door Knocker

THE DEVIL'S PACT

There was a man so envious
His neighbour fair and tall
He burnt with hate and foul desire
And vowed to have it all

The Devil lured, a pact was made
To sacrifice his son
His neighbour's limbs to change for his
And quickly it was done

Happy was he for just a while
Then envious as before
He burnt with hate and foul desire
And soon he wanted more

The Devil lured, a pact was made
His daughter he did yield
His neighbour's body was exchanged
To serve the lust he willed

Content was he for just a while
Then envious as before
He burnt with hate and foul desire
And yet he wanted more

Ficus Tree Roots

The Devil lured, a pact was made
He sacrificed his wife
His neighbour's looks exchanged for his
And thus was his for life

Happy was he for just a while
Then envious as before
He burnt with hate and foul desire
And still he wanted more

The Devil lured, a pact was made
He offered up his soul
His neighbour's thoughts and consciousness
Transferred to him in whole

Then in a flash he realised
The folly of his act
He had become his self again
Beware the Devil's pact!

Buddhist Temples

REFLECTIONS OF BUDDHISM

By chance encounter
In life's passing dream
Do for the doing
With deep compassion
Yet non-attachment

Earthly vanities
All but illusions
Mere shadows bubbles
Impermanent as
Sun-touched morning dew
Or flash of lightning
Through the emptiness
Of time's endless void

To reach the Pure Land
Seek the right path from
Neither shrine nor scroll
True enlightenment
Is found within you

Part V

Last Words

Common Corncockle

PRAYER

O Lord to Whom we all pray
Make us snow-pure by Thy grace
Keep out evil from our days
Lead us through life all the way

Lead us on the path of truth
Let us follow virtue's way
Walk with beauty every day
Thus to find eternal youth

May our spirit never age
But retain its innocence
And by Thy munificence
Record a meaningful page

In the book of life we write
Labour laughter joy and pain
We shall not return again
Make it worthy of Thy sight

From this dusty world we pass
Into Thy arms like the dew
That Thou lift into the blue
Lightly from the morning grass

Statue at a Mausoleum

PIETY

Ere the searching soul should hear
Destiny's mournful knell
Piety comforts our fear
Of death's eternal spell
We make our own heaven here
Or suffer our own hell

Give the heart to godliness
Tied with a pious cord
Raise the arm of righteousness
Bury the hateful sword
Be a well of happiness
Walk humbly with the Lord

Statue of Eve by Rodin

FAREWELL THE PAST

Never idealize the past
You will not find life there
Worldly vanities do not last
And only a fool would care

Never journey into the past
You will be a stranger there
Images of youth are fading fast
That your vulnerable soul laid bare

Never go romancing the past
You will not find love there
The crucible where the heart was cast
Endures despite the wear

Never linger in the past
You will not find your way there
The promises of life however vast
Are only for those who dare

Snow Covered Bare Tree Trunk

PARTING IN WINTER

It is winter
When sobering snow flies
From a desolate sky
Into white emptiness

It is hopelessness
Of sunken cheeks and sightless eyes
When life vainly strains
For every gasping breath

It is sadness
For a departing soul
Whose dreams linger
Though breath is no more

It is warm tears
From a bursting heart
When silent grief
Wells from within

In Search of Beauty

Winter Scene

It is freedom
For a cleansed spirit
To walk the icy earth
Comforted by falling snow

It is a tomb
For well earned rest
Where faith fosters
Eternal peace

Three Yellow Tulips

REQUIEM FOR V.Y.C.C.

Death, prepare a friendly tomb
Take his hand and lead him home
Though his mortal frame may part
Save the honour of his heart

An offering this skinny bag
Though these wholesome features sag
Wearing your inglorious mask
Leaving an unfinished task

This and Heaven's promised gate
Ease acceptance of man's fate –
Immortality's decree:
'What has been shall always be'

Ere the all-conquering worm
Eats the sinew of his term
God have mercy on his art
Make a small place in Thy heart

Make a blessing of our days
Set the willing heart ablaze
Steadfast in the timeless night
Make a beacon of his light

Hare Statue by Flanagan

IN MEMORIAM T.A.B.

The net of death was cast afield
It gripped you in your prime
Your shining hopes by darkness sealed
In sacrifice to time
Just yesterday your smiles would flash
A face so young and kind
Tonight an epitaph of ash
Is scattered by the wind

The woodland turned from green to gold
And leaf by sad leaf fell
There you lay friendless in the cold
Without a last farewell
The flame of life a tender light
The splendour of a star
Now faded in this cheerless night
Remembrance is a scar

Sunset behind Palm Tree

The dream of youth a fantasy
Its brief enchantment gives
At times a taste of ecstasy
And in our hearts it lives
Each passing day for us enweaves
The drama of the age
And each departed player leaves
Faint echoes on the stage

The future holds in mystery
The promise of the dawn
Each evening sets in history
The fair day it has drawn
The present teaches us no fear
Will bear us to the last
Then firmly and without a tear
We walk into the past

Cloud Reflection

TO VICTOR

What do I remember you by
Not sadness in a parting sigh
But beauty by your artist eye
And colours from your darkened sky

What do I remember you by
Not inward tears that cannot dry
But the sound of your hearty cry
And sharing laughter you and I

What do I remember you by
Not heartache of a fond good-bye
But freedom from this earthly tie
And peace under a different sky

What do I remember you by
Not the mystery as to why
But our destiny to defy
Through aspirations noble high

Courage worn on a hopeless face
Brought sweetness to a last embrace
Ever cherished your heart and grace
Everlasting in time and space

Buttercups

JOURNEY TO THE STARS

When I leave this world behind me
On a journey to distant stars
Towards an unknown galaxy
Far beyond familiar Mars

I bear wonderful memories
Of the beauty of Mother Earth
From the shy night's dark mysteries
To the proud day's colourful birth

I bear heart-warming memories
Of the goodness of Mother Earth
The challenge of humanity's
Bold resurrection and rebirth

These thoughts shall accompany me
For the journey will never end
To travel through eternity
With the Creator as a friend

Daffodil

LAST SONNET

Seek ecstasy where mist-enshrouded Spring
Stirs up the pregnant earth by her caress
To bring a living green to everything
Then looks with envy at their loveliness
There clad in beauty rise the virgin hills
Where waves the verdant tress of slender trees
Far in the vale amongst the daffodils
In running ripples roams the rambling breeze
For though our place in time but empty years
When folly firmly holds the helm of life
Our cowardice has nourished all our fears
Our weakness triumphs in our inner strife
Yet in my dying rattle halting breath
Still in this beauty shall I outlive death

Appendix

Full Moon

夜 思

牀 前 明 月 光
疑 是 地 上 霜
舉 頭 望 明 月
低 頭 思 故 鄉

by Li Bai（李白）*701 A.D. – 762 A.D.*

(Translation)

NIGHT REMINISCENCE

Blazing moonlight by my bed
Shines like frost upon the town
Bright moon rouses up my head
Tender home thoughts weigh it down